MOLLUSKS

Rebecca Woodbury, Ph.D., M.Ed.

Gravitas Publications Inc.

MOLLUSKS

Illustrations: Janet Moneymaker

Mollusks
ISBN 978-1-950415-59-5

Published by Gravitas Publications Inc.
Imprint: Real Science-4-Kids
www.gravitaspublications.com
www.realscience4kids.com

RS4K

Photo credits: Cover & Title Pg: cherylvb, AdobeStock; Above, macrophile on Flickr, CC BY SA 2.0; P.5: 1. Albert Kok, Public Domain; 2. Andreas Eichler, CC BY SA 3.0; 3. NOAA Okeanos Explorer Program, INDEX-SATAL 2010; 4. Andrew David, NOAA/NMFS/SEFSC Panama City, Lance Horn, UNCW/NURC-Phantom II ROV operator; P.7: 1. Jess Van Dyke, Snail Busters, LLC, Bugwood.org; 2. Paul Asman and Jill Lenoble, CC BY SA 2.0; 3. Manfred Richter from Pixabay; P.8: 1. Guillaume Brocker, CC BY SA 3.0; 2. Claire Fackler, CINMS, NOAA; 3. Betty Wills, CC BY SA 4.0; P.9: 1. David Burdic, NOAA; 2. NURC/UNCW and NOAA/FGBNMS; 3. Tango22, CC BY SA 3.0; P.11: 1. Guillaume Brocker, CC BY SA 3.0; 2. Waugsberg, CC BY SA 3.0; 3. NOAA Okeanos Explorer Program, INDEX-SATAL 2010; 4. David Monniaux, CC BY SA 3.0; 5. © Hans Hillewaert-CC BY SA 4.0; 6. Octopus, NURC/UNCW and NOAA/FGBNMS; P.16. Oyster, Nick Hobgood, CC BY SA 3.0; P.17. Nick Hobgood, CC BY SA 3.0; P.18: 1. © Hans Hillewaert-CC BY SA 4.0; 2. Nick Hobgood, CC BY SA 3.0; 3. L. Shyamal, CC BY SA 3.0; P.19. Waugsberg, CC BY SA 3.0; P.20: Left, Radula Illustration, Debivort at en.wikipedia, CC BY SA 3.0; Right, Radula micrograph, Robert Hershler & Hsiu-Ping Liu, CC BY SA 3.0

Have you ever eaten **snails, clams, squid,** or **octopus** for dinner?

Just give me cheese, please!

Snails, clams, squids, and octopuses are called **mollusks.**

Octopus

1

Snail

2

Clam

3

Squid

4

Mollusks are soft-bodied
animals that live in oceans
and lakes and on land.

I see mollusks
in my garden!

Some mollusks live in lakes and ponds.

Some mollusks
live in oceans.

Some live
on land.

Mollusks look very different from one another.

Slug

Nudibranch

Squid

Clam

Octopus

Limpets

All mollusks are in the group called **Mollusca.**

There are three different types of mollusks in the group Mollusca.

Gastropods ┈┈➤ Snails and slugs

Bivalves ┈┈┈➤ Clams and oysters

Cephalopods ┈➤ Octopuses and squids

Gastropods

Slug

1

Snail

2

Bivalves

Clam

3

4

Oyster

Cephalopods

Squid

5

Octopus

6

All mollusks have the
same **basic body plan.**

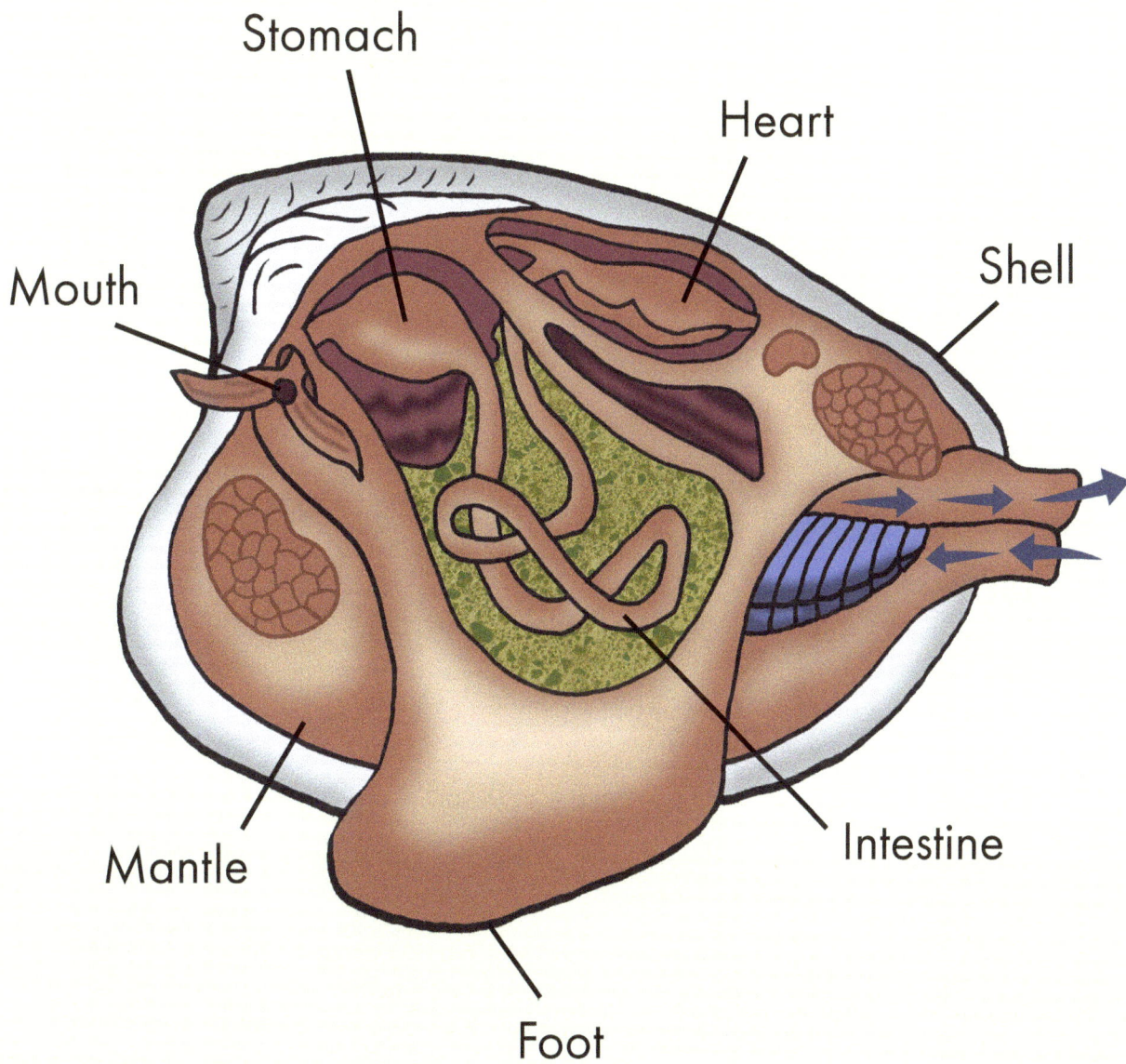

Stomach

Heart

Shell

Mouth

Mantle

Intestine

Foot

All mollusks have **organs.**

An organ is a body part that has a particular job to do in the body.

The **heart, stomach,** and **intestines** are organs.

We have those organs too!

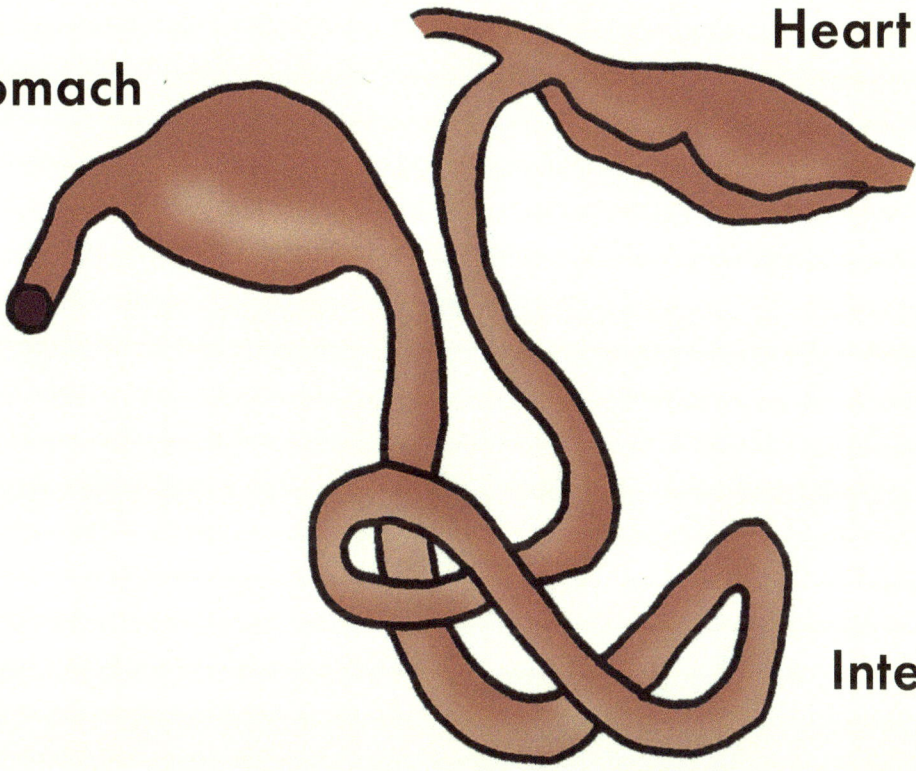

Stomach

Heart

Intestine

Most mollusks have a shell.

Some shells cover the body of the mollusk.

Oyster

Some shells are small
and are inside the mollusk.

Squid

A **mollusk** has a **mantle**.

The mantle covers and protects the organs and creates the shell.

The colorful part inside the clam's shell is part of the mantle.

Behind the head of the squid, you can see the outside of the mantle.

The white part sticking out from the snail's shell is part of the mantle.

A mollusk has one or more **feet.** A foot is a large muscle used for moving.

Foot

Many mollusks have a toothed tongue called a **radula.**

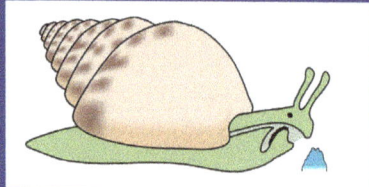

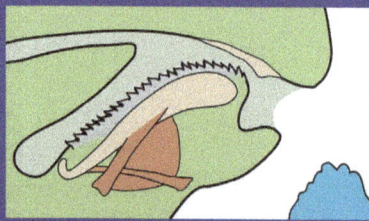

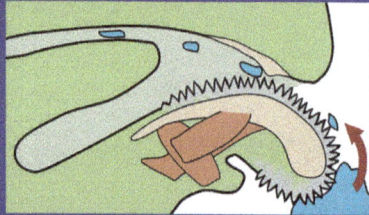

A snail uses its radula to scrape up food.

Teeth on a radula seen with a microscope

Mollusca is a very big group of animals that come in many different types and many different shapes, sizes, and colors.

Mollusks are everywhere!

How to say science words

bivalve   (BIY-vaalv)

cephalopod   (SEH-fuh-luh-pahd)

clam   (KLAAM)

foot   (FUHT)

gastropod   (GAA-struh-pahd)

heart   (HAHRT)

intestine   (in-TEH-stuhn)

limpet   (LIM-puht)

mantle   (MAAN-tuhl)

Mollusca   (mah-LUH-skuh)

mollusk   (MAH-luhsk)

muscle   (MUH-suhl)

nudibranch   (NOO-duh-braank)

octopus   (AHK-tuh-puhs)

organ   (AWR-guhn)

oyster   (OY-stuhr)

radula   (RAA-juh-luh)

science   (SIY-uhns)

slug   (SLUHG)

snail   (SNAYL)

squid   (SKWID)

stomach   (STUH-muhk)

tongue   (TUHNG)

www.ingramcontent.com/pod-product-compliance
Lightning Source LLC
Chambersburg PA
CBHW040150200326
41520CB00028B/7561